DRIVE-IN DELUXE

Michael Karl Witzel

Motorbooks International
Publishers & Wholesalers ®

Dedication

This book is dedicated to the fully operational, no-excuse curb service, dining-in-your-car-type drive-in restaurants that have made a fortuitous return to the American scene, along with the gradual rediscovery of improved fast food quality and friendly customer service. America's grill workers, carhops, waiters, waitresses, and cashiers: Keep up the good work and don't let the unchecked proliferation of the fast food franchises get you down. Delicious food prepared well and served up with style will keep the car customers coming back again and again.

First published in 1997 by Motorbooks International Publishers & Wholesalers, 729 Prospect Avenue, PO Box 1, Osceola, WI 54020-0001

Motorbooks International is a certified trademark, registered with the United States Patent Office

Motorbooks International books are also available at discounts in bulk quantity for industrial or sales-promotional use. For details write to Special Sales Manager at the Publisher's address

Library of Congress Cataloging-in-Publication Available
ISBN 0-7603-0211-1

On the front cover: Good tunes and good food, the classic drive-in combination. *Michael Karl Witzel*

On the frontispiece: The ceaseless homogenization of America marches on leaving many independent drive-ins and roadside joints in the dust. *Michael Karl Witzel*

On the title page: A & W Drive-Ins were a big part of the 1950s in many parts of the country. Starting out as drive-in diners that used carhops to serve customers, modernized units of the postwar era installed electronic speaker box systems to aid order taking. *A & W Restaurants, Inc.*

On the back cover: Matchbooks were at one time given away free by the majority of businesses in America. Today, they are hoarded by drive-in memorabilia collectors and join similar articles as reminders of long-defunct operations. *Michael Karl Witzel*

Printed in Hong Kong

CONTENTS

Acknowledgments

A triple-thick thank you goes out to all the people who donated time, knowledge, recollections, photos, advertisements, contacts, and suggestions, most notably A & W Restaurants, Inc.; the Coca-Cola Company; Michael Dregni; the Dr. Pepper Museum of Waco, Texas; Albert Doumar; Ralph Grossman; Hanna-Barbera Entertainment Company; Ramona Longpre; Marriott Corporation; Richard McDonald; Clare Patterson; Louise Sivils; Sonic Industries, Inc.; Steak n' Shake, Inc.; Texas Pig Stands Inc.; Universal Studios Florida; Buna "Johnnie" Van Hekken; Steven Weiss; and June Wian. A steaming-hot pot of fresh coffee is reserved for the many archives, artists, and talented photographers that provided appropriate drive-in images, including Howard Ande, the Atlanta Historical Society, Kent Bash, Annabelle Breakey, the helpful and always knowledgeable staff at CoolStock, the great researchers at the Dallas Public Library, the Georgia Historical Society, Glen Icanberry, Brett Parker, Louis Persat, Preziosi Postcards, Quincy Historical Society, Jim Ross, the Security Pacific Historical Photograph Collection, Bob Sigmon, Andy Southard, Jr., Tombrock Corporation, University of Louisville Ekstrom Library, and Randy Welborn. Finally, a second helping of thanks is served to Joan Johnson of Circa Research and Reference, Douglas Photographic, and Imagers.

The author hanging out at Mel's Drive-In restaurant, seen here with the classic, yellow deuce coupe of *American Graffiti* movie fame and the 1950 custom Mercury "MRL MIST." The hot rod is now owned by car collector Rick Figari of San Francisco, California, and the lead sled by Marvin Giambastiani. *Annabelle Breakey*

WHAT IS A DRIVE-IN RESTAURANT?

According to strict definition, a drive-in restaurant is a specialized eating establishment that employs a staff of parking lot servers, or carhops, to deliver food and drinks to customers waiting in their cars. There, a small tray that holds the ordered items is clipped onto the window where it provides a miniature table for the dining visitor. As they eat, patrons may enjoy private conversation in the car, listen to the radio, or just sit back, relax, and watch the sights. At any time, they may make an additional order or request from the carhop on call. When their appetites are satiated, it's customary that they leave a small tip on the tray. When they are ready to leave, the tray is removed from the car, and they may pull out into traffic unhindered.

The term *drive-in* has grown to include a much wider range of roadside restaurants and now may be applied to dining spots that don't offer carhop service. But to gain admittance into the wayside fraternity of the traditional drive-in restaurant, there are a few rules that must be followed: First, the eatery must be visited primarily by patrons arriving by automobile, motorcycle, motor home, or other type of conveyance (food stands at the local mall are definitely out). The drive-in must serve up delicious, old-fashioned drinks, desserts, and road food and be free of all of the pomp and circumstance that's found at the more upscale sit-down restaurants. Most important, it must convey the ambiance so adored by car-food connoisseurs.

In the visual department, vintage architecture is a big plus, as is a prodigious amount of neon tubing, the availability of tabletop jukebox controls (with three plays for a quarter) in those establishments that offer indoor seating, and menus that are encased in thick plastic with those little metal tabs on the corners to protect them from wear and tear. The reticent waitress who's wearing a pair of those funny little cat's-eye glasses (preferably attached to a chain) and a classic pink uniform with her first name embroidered on the front pocket can often bring an ordinary greasy spoon into the hallowed category of the drive-in. It's sufficient to say that any eatery offering the slightest hint of nostalgia—and boasting a heritage in pop culture—fits into the category of "drive-in deluxe."

With this broader definition in mind, your nearest drive-in restaurant could very well be the gleaming stainless steel diner parked down the boulevard. At the same time, the funky little coffee shop that's located downtown—the one where everyone hangs out sipping coffee and smoking cigarettes—could be considered a drive-in as well. When the craving for creamy dessert foods arises, it's the walk-up dairy stand located on the outskirts of town that provides the drive-in satisfaction so many people are looking for. In that same vein, the myriad walk-up burger joints that cling to survival all across America are drive-ins in their own right, as are the tiny mom-and-pop taco shops, hot dog havens, burrito bars, and soft drink stands that make up that fast food underbelly of the culinary world. The American drive-in restaurant is a state of mind, a feeling that's waiting for us out there along the back roads of America, usually when we least expect it and our taste buds are ready for a roadside treat.

Chapter One

THE ORIGINS OF DRIVE-IN DINING

*D*uring the latter part of the 1800s, most of America's city pharmacies housed the bubbling apparatus used to produce carbonated water. In the sunny South, hot summer temperatures heated up the growing demand for syrup-flavored soda until it became one of the standard commodities of Main Street. By the turn of the century and the dawn of the motorcar, the obsession to consume soda water became the catalyst that forced conventional food and drink operations to consider new and improved ways of serving customers.

To satisfy the growing demand for those tasty soda-water treats that tickled the taste buds and soothed the palate, savvy merchandisers established a new precedent for public refreshment and diversion. Much like the ubiquitous fast food restaurants of the present day, a myriad of commercial enterprises sprang up along the urban

In the late part of the second decade, the Edgewood Pharmacy in Dallas, Texas, was typical of similar drugstore operations doing business. Runners were employed as waiters to the curbside, taking out to waiting cars refreshments that were previously available only to walk-in customers. The format was the predecessor to the drive-ins that emerged during the early 1920s. *Dallas Public Library*

9

The drive-in restaurant made its debut in the Hanna-Barbera television cartoon the *Flintstones*. Back in those prehistoric times, a big rack of ribs was a popular item, in spite of the fact that its sheer weight could tip over the vehicle. *Hanna-Barbera Productions, Inc. ©1996, All Rights Reserved.*
Turner Broadcasting Systems, Inc. ©1996, All Rights Reserved

corridors in order to take advantage of the burgeoning trend. The city of Atlanta, Georgia, was a typical example of the increasing availability of fizzy liquid fun: By 1886, five fully equipped soda fountains were dispensing their sweet delights throughout the town and even more fountains were on the way!

As citizens made visiting the local pharmacy and its soda bar a recognized part of their daily rituals, the act of consuming nonalcoholic beverages simply for pleasure gained increased acceptance as a social activity. Arriving by foot, horse, or carriage—both men, women, and children regularly frequented the soda fountains in an effort to relax and escape their everyday woes. In an era that was devoid of radio, television, and other forms of popular entertainment, sipping a soda was found to be an excellent way to pass the time and to socialize with others (the future patrons of the drive-in eateries would mirror this same cultural activity).

At the turn of the century, the Coca-Cola Company promoted its syrup throughout the South with posters and point-of-purchase items that depicted people in their carriages being served by waiters. These runners were often employed by pharmacies to carry out refreshing beverages to people waiting in their horse-drawn coaches.
Courtesy Coca-Cola Company

As more and more pharmacies decided to add car service to their way of doing business, the moveable serving tray became a part of the everyday utensils used to serve the public. Trays attached to the window or door and featured some sort of bracket that braced against the body of the automobile.
Dallas Public Library

By the late 1920s, the Texas Pig Stands were moving away from their simple beginnings and began to try new forms of architecture for their drive-in units. This simple, octagonal concession design (constructed in the city of Los Angeles) was a precursor to the grand monuments that were erected by Harry Carpenter and Bill Simon during the 1930s. *Courtesy Texas Pig Stands, Inc.*

Crammed in among the pills and potions, the typical soda fountain serving arrangement included a large counter area or elaborate marble bar that functioned as a concession stand. Some establishments even provided areas where ornate tables and chairs were set up for more relaxed drinking. But as civilized as these conditions were, the growing penchant for consuming tasty drink invariably caused some problems. The menfolk, who had a nasty habit of chewing tobacco, smoking cigars, and engaging in bawdy conversation, often offended the more genteel female clientele. When the local soda fountains became overly crowded, sensibilities were assaulted.

The indignities weren't ignored for long and a number of solutions were examined. One of the most memorable accommodations to be made

The Texas Pig Stand operation is credited with being the first in America to implement the idea of a dedicated drive-in that serves food and drinks. Their present-day restaurant—one of many—still operates in San Antonio, Texas, and pays homage to the early years with this sign posted at the carhop lanes. *Michael Karl Witzel*

Steak n' Shake restaurants were started by Gus Belt back in 1934. He based the operation on a "Steakburger" that was made with finer cuts of ground beef and milkshakes that were hand-dipped with real ice cream. Although the carhops are ancient history, the outfit still thrives as a popular fast food chain throughout the midwestern regions of the United States. *Howard Ande*

for the express convenience of "mobile" customers gained notoriety in Memphis, Tennessee. It all started in 1905 when horse-drawn carriages were pulling up to Harold Fortune's drugstore carrying passengers eager to partake of the tasty potions sold inside. According to legend, the business was so brisk one summer evening that a local man came up with the revolutionary idea of allowing the ladies to wait in their carriages while he braved the crowds and made all the drink orders. He then hand-delivered the sarsaparilla and other concoctions to the women as they perched in their surreys.

Of course this gentlemanly style of outside service was received quite well by all who were there, and by the end of the summer, word of the incident had spread throughout the town. Sud-

denly, upscale patrons were driving by with their chauffeured carriages to take advantage of the new "service." Caught off guard, Mr. Fortune was forced to rely on the young lads who were already employed at his fountain as counter servers. Unfortunately, it was soon discovered that the new procedure took too much time away from the employee's normal activities. A staff of delivery boys was hired to assume the hurried task of delivering soda water to people queuing up out in the street.

The setup became so immensely popular that even the Coca-Cola Company, one of the leading producers of syrup for the fountains, adopted the idea of serving people in their carriages as the theme for one of their early print advertising campaigns. Posters showing fashion-

The Texas Pig Stands experimented with a number of building styles during the drive-in's formative years. One of their most unusual designs was a pig-shaped structure that operated in the south-central Texas area. Recently, the vintage stand was discovered and plans are currently in work to restore the structure and return it to its former glory. *Michael Karl Witzel*

The Pig n' Whistle was one of the classic roadside eateries that opened for business in the South. The chain began in California as a tearoom in 1910 and spread across the country until it stopped in Macon, Georgia. Drive-in service was added in the 1920s, and by the 1950s future celebrities such as Little Richard, Otis Redding, and James Brown worked there as carhops. A sandwich called the "Pig Special" once sold there for 20 cents. "That piping, dancing little Pig with the jolly, 'come hither' smile has all sorts of good things for your refreshment and entertainment. . . ." *Author collection*

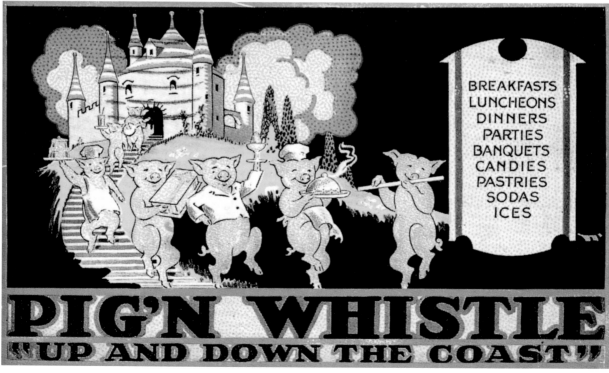

BREAKFASTS
LUNCHEONS
DINNERS
PARTIES
BANQUETS
CANDIES
PASTRIES
SODAS
ICES

PIG'N WHISTLE
"UP AND DOWN THE COAST"

As the pharmacy gained notoriety as the place to hang out and take a cool soda refreshment, the lad working behind the counter attained a certain mystery and mystique about him. Known by the public as a "soda jerk," this nonalcoholic barkeep appeared in numerous movies and slowly worked his way into the popular culture of our nation. *CoolStock/Library of Congress photo*

able women drinking their refreshments in their carriages caught the public's eye and furthered the cause of curbside soda fountain service.

After the automobile gained a strong foothold in the domestic market, the imbibe-in-your-vehicle concept was taken to an entirely new level: revised ads portrayed a carload of revelers enjoying effervescent tumblers of Coca-Cola while seated in their motoring coach. If that wasn't convenient enough, additional point-of-purchase prints showcased waiters who delivered beverages out to trains, boats, and other modes of transportation.

By that time, Fortune had relocated his growing fountain operation to the busy business district of Memphis and was introducing neophyte motoring enthusiasts to the magic of car service. Business became so good that a rabble of overeager automobilists frequently jammed the streets in an effort to get a cool drink. A hasty plan was implemented that called for the curb to be taken out in front of the establishment so that the waiting vehicles could pull up more closely to the front of the building (and right up on the sidewalk). Unfortunately, the stop-gap measure failed to alleviate the car congestion.

The Memphis city fathers overreacted to the complaints of nearby businesses and enacted an ordinance that banned any sort of car-oriented curb service in the downtown business district. Once again, Fortune moved his outfit—this time to a remote location on the outskirts of the city where traffic control and the complaints of neighbors wouldn't be a problem. It was 1922 when he constructed a brand new store that featured an even larger, separate fountain operation, kitchen facilities, and an oversized parking lot area. At last, dozens of automobiles could pull right in, find a space, and take advantage of unhurried front-seat service.

Surprisingly, Fortune was not the first to open the nation's premiere dedicated drive-in restaurant. One year earlier, the visionary team of Dallas, Texas, candy and tobacco magnate Jessie G. Kirby and physician Reuben W. Jackson beat Fortune at the drive-in dining game by opening the first roadside eatery that was specifically earmarked to sell lunch, dinner, and bottled refreshments at the curb. Unlike Fortune's Memphis fountain operation and the dozens of other imitators who copied its success throughout the South, the Texas restaurant was designed and built from the ground up with its sole intention to serve modern motorists right in their cars. This devoted drive-in concept was the brainchild of Kirby who had a hunch that the habits of the lackadaisical vehicle owner could be readily converted into cold, hard cash. "People with cars are so lazy that they don't want to get out of them to eat!" was the creed that he uttered when he tried to convince Jackson,

his future partner, to invest the capital Kirby needed to start up the nation's first drive-in sandwich business. Jackson was captivated by Kirby's dramatic sales pitch and provided the $10,000 financing needed to build the prototype restaurant stand that was geared specifically to serving patrons while they remained seated in their cars.

In the fall of 1921, two American icons—the roadside restaurant and the motorcar—gave birth to an entirely new form of dining, the drive-in restaurant. That year, Kirby and Jackson's ambitious eatery, dubbed the "Pig Stand," began selling its delicious fare of "Pig sandwiches" and soft drinks from a small shack on the busy Dallas-Fort Worth Highway. Kirby's initial hunch was correct, and before too long the Lone Star state was abuzz with complimentary news of the Pig Stand's friendly carhop service, great food, and superior convenience. The drive-in was king.

Thirty-four years later both Kirby and Jackson had passed on, but Pig Stands were going stronger than ever. Former carhop Royce Hailey took over the reins and led the eatery and the numerous duplicates that bore its name into the halcyon days of curbside car service. By that time, the once novel idea of the "new motor lunch" had become part and parcel of growing up in America and traveling by car. With an endless parade of imitators, the likes of Pig n' Whistle, Marriott's, Mel's, and many others, carrying on the drive-in dream from coast-to-coast, the drive-in restaurant was poised to remain the foundation of roadside commerce for years to come.

Made famous in the George Lucas film *American Graffiti*, the Mel's Drive-In that was located on San Francisco's South Van Ness was truly the quintessential curb service operation. Today, Steven Weiss continues to operate Mel's units throughout California and to provide the nostalgic atmosphere of the 1950s to modern-day customers. *CoolStock/Steven Weiss photo*

Chapter Two

CARHOPS AND CUSTOMER SERVICE

*B*y the time America's drive-in restaurants reached their zenith in the 1950s, the parking lot server had evolved to become an integral part of the ambiance and pop culture of roadside dining. Eating out in your car meant personal interaction with a carhop—a living, breathing entity with feelings and emotions that catered to every whim when it came to food and drink service along the road.

As we learned in the previous chapter, the story of the first carhops began at the turn of the century when the young lads who were employed at pharmacy soda fountains doubled as deliverymen to carriage-mounted customers at curbside.

As the decades progressed, a small number of businesses experimented with this new format, albeit in a limited fashion. It wasn't until 1921 that the fantasy of being served in one's vehicle by a

The ordering board is used by Sonic drive-ins as a way to show exactly what food and beverage items are available on the menu. The customer depresses a button and addresses a worker inside the building who takes his order. Compared to the other fast food menu boards currently in use, it's a setup that allows the patron some semblance of control.
Michael Karl Witzel

19

Kim's drive-in is a landmark for car service in Waco, Texas. The town boasts a rich history of drive-in eateries that at one time featured car service by way of carhops. Today, the elaborate Las Vegas-style sign that attracted customers back in the 1950s still shines. *Michael Karl Witzel*

The McDonnell's drive-ins operated in the Los Angeles area circa 1930. The owner, Rusty McDonnell, employed the services of a local artist to sketch whimsical representations of the typical male and female carhop on his menus. Back then paper menus were a large part of dining in a motorcar and took the place of today's impersonal, illuminated menu boards. *Author collection*

waiter gained any great notoriety. By then, the motorcar had entrenched itself in virtually every level of society, and a bond between man and machine had formed. Sure, cars were practical means for getting to and from work, but they also provided one limitless hours of pleasure. Why not consume one's lunch or dinner right behind the steering wheel?

Texas' Pig Stand was the prototype drive-in. With its construction in 1921, a brand new protocol for customer interaction was introduced. As customers in their automobiles pulled up to the curb, young men assigned to serve them eagerly jumped up onto the running boards of the cars before the vehicles even came to a stop! Orders were taken and relayed by foot to the cook shack where the food was prepared. When ready, the barbecued fare—"Pig sandwiches"—were hustled to the customer where they were consumed in the front seat of the automobile. For the busy carhops, speed and agility were of the utmost importance since the wages of drive-in servers in those days consisted entirely of customer tips.

It was this rather aggressive style of automobile service that influenced the creation of the term "carhop" and the widespread acceptance of in-car dining. As more and more drive-in eateries appeared to copy the Pig Stand's great success, Americans were finding out that their cars were ideal facilitators for all sorts of fun and frolic. By the end of the 1920s, drive-ins were well on their

Lorraine Magowan worked as a carhop at the Los Angeles Simon's Drive-In during the late 1930s. At the time most of the curb gals working in the area borrowed their fashion cues from the military and donned the basic hat, slacks, and simple shirt arrangement seen here. *Courtesy Lorraine Magowan*

During the second decade of this century, the carhop, or runner as he was then called, was almost always male. Pharmacies (like this one in Dallas, Texas) employed the services of the soda jerk and later hired dedicated delivery boys to bring liquid refreshments out to the cars. After the majority of men went off to fight in various wars, more women entered the workforce and eventually serving automobiles at the drive-in became the domain of the female. *Dallas Public Library*

way to becoming part of this nation's popular culture. Among the mobile cognoscenti, it was acknowledged that eating in your car with the top down was the preferable way to dine. The high-speed delivery of the always flamboyant carhop readily satisfied the growing demand.

Meanwhile Roy Allen, who was operating a small root beer stand near Lodi, California, began experimenting with the next phase of in-car service—what he called "tray girls." Upon discovering that female servers caused an increase in business, Allen added even more to his staff and watched as his operations grew. By the mid-1930s his A & W Root Beer brand was a household name and throngs of customers were driving out onto the roadways in search of a frosty mug. Of course, some male car owners had ulterior motives: They were venturing out to see the new legion of female carhops strut their stuff.

Since many of the drive-ins across the country were switching to female carhops, that was relatively easy to do. World War II exacerbated the situation as much of the male workforce was shipped overseas leaving women to take over the existing carhop jobs. Before too long, wages at the local drive-in were looking almost as good as those pulled down by "Rosie the Riveter" at the bomber plants in southern California. To earn some extra cash, free hours were soon spent at

Carl's drive-ins of southern California employed carhops who wore elaborate outfits a bit different from the competition. Instead of the usual slack, shirt, and jaunty cap combination that was so common at the time, hops at Carl's were required to wear ruffled skirts. *Security Pacific Historical Photograph Collection*

During the 1950s, Ramona Longpre (center) was the carhop's carhop at Mel's classic drive-in eatery in Berkeley, California. Pictured here in various working situations with her coworkers, the top right image highlights a reunion the curb gals had back in 1988 (from left: Polly, Ramona Longpre, Mary Carrico, and Ollie). All four worked as carhops in the Berkeley area during the heyday of drive-ins. *CoolStock/Ramona Longpre photos*

the local drive-in burger joint jotting down food orders and slinging trays.

Naturally, restaurant proprietors were pleasantly surprised by just how much additional business could be pulled in with female help—especially when that help was outfitted in a revealing costume. Sure, a natty curb boy outfitted with white shirt and black bow tie projected a nice, professional image, but he didn't have nearly the impact of a waitress who sauntered around a parking lot wearing what some publications of the day referred to as "silk pajamas"! That kind of image could really pull 'em in.

One of the most visible developments regarding American carhop fashion occurred in 1938 when J. D. and Louise Sivils opened an elaborate "drive-in" operation in the city of Houston,

Texas. Louise knew exactly what she wanted and her first order of business was to create eye-popping uniforms. Taking inspiration from a series of Chesterfield cigarette ads, she dressed her girls in revamped majorette outfits. An able seamstress, she made the outfits herself and adopted shimmery satin as the material of choice when it came to clothing her "curb girls."

The Sivils garb set the new standard for curb service. Comprised of a tightly fitted majorette-style jacket teamed up with a matching pair of shorts, the set was distinguished by an elaborate plumed hat. Keeping the locality in mind, fancy cowboy boots were used for footwear (new girls

Marbett's was a drive-in restaurant outfit that was a spin-off of the famous White Tower hamburger chain. Their gleaming, porcelain clad, Streamline Moderne buildings were premanufactured by the Valentine Diner Company of Wichita, Kansas. Here, carhops led by their male leader (center) assemble prior to the day's service. *CoolStock/Photo courtesy Tombrock Corporation*

were required to buy their own). Although they made the gals look great, the decorative outfits definitely had drawbacks. During the hot summer months the combination of high-top boots and nonbreathable satin turned many a little girl's dream of car service into a nightmare.

Despite the often uncomfortable working apparel, the allure of becoming a carhop became ever greater. Aided by imaginative operators who turned the simple act of hanging a tray load of hamburgers, Cokes, French fries, and milk shakes on a parked car into a new form of American theater, the drive-in mystique spread like wildfire throughout the country. Suddenly, everyone wanted to get a job at the local drive-in eatery and the position of carhop attained as much status as the job of airline stewardess or nurse. By the time our best and brightest were off to fight a new war

America's choice

You and your family will always find a friendly welcome at Howard Johnson's . . . a pleasant atmosphere . . . good food at sensible prices served by a waitress trained to bring you the best in courteous, efficient, friendly service . . . no matter where you travel!

HOWARD JOHNSON'S
RESTAURANTS · MOTOR LODGES
Ice Cream · Candies · Take-Home Frozen Foods
"Landmark for Hungry Americans"

Howard Johnson's used to be one of the most popular dining stops to drive into when taking a trip. Although she didn't provide car service, the iconic image of the American waitress was a major selling point for the restaurant chain, as was the 28 flavors of ice cream available within.
Author collection

in Korea, America's carhops were truly the "belles of the boulevard."

After carhop Josephine Powell appeared on the cover of *Life* magazine, the longing for curb jobs turned to a wild frenzy. With all the future beauty queens and debutantes high-stepping around on American parking lots, the drive-in restaurant definitely became the place for those in cars to sit back and watch the show. Suffice it to say that there was many a male hot-rodder who sat in his roadster slowly sipping at a milk shake while dreaming of a date with one of these service celebrities of the strip.

By the end of the 1950s, the stereotypical carhop of the type depicted in the movie *American Graffiti* made her debut at the drive-in. Now she was clad in tight-fitting spandex slacks with a white blouse and jaunty pillbox hat. She had an attitude to match and took full advantage of her newly acquired status. After roller skates were strapped to her feet to make operations move a bit faster, her image transcended from the able car-servant that she was to a full-blown American icon. The car-crazy public that visited drive-ins and wolfed down food in their shiny new automobiles knew what they liked, and this was it.

But in the long run, even the extra speed afforded by roller skates couldn't save carhop service. As the 1960s turned to the 1970s, the pace of life clicked up yet another notch and the public demanded even faster food. Multiple trips made by carhops to take the order, fill it, and return for the tray were viewed as redundant. At the same time, real-estate and other operating costs were rising. To save money, the nation's drive-ins clamored for methods to free them from human tray carriers.

Eventually, speaker box systems using vacuum tubes and miles of wiring replaced carhops, until finally she was the dinosaur of the drive-in trade. A pole-mounted intercom setup superseded all of the ordering duties formerly done in person. A person sitting behind the "drive-thru" window assumed the responsibility of handing out the entrees. With that, the personal touch was gone. The public had come to accept the utility of impersonal gadgetry and had forgotten what it was like to be served by a real person. The era of carhops and customer service was over.

The carhop was a frequent visitor in the print advertisements that were used by the drive-ins down through the ages. Here, the newly improved version of the 1960s A & W carhop has been remade as a streamlined reincarnation of her predecessor, now donning a modern cap that would look at home on any jetliner stewardess or military employee. *Author collection/courtesy A & W Restaurants, Inc.*

Thirst-aid station

A & W ROOT BEER...brewed with pure natural ingredients and true draft flavor ...so good with food!

Fun and refreshments are always ready for you at the 2400 A&W "thirst aid stations" coast-to-coast and overseas. A stop for A&W's delicious food and drinks makes a hit with everyone — even baby (under 5 years) gets a FREE Baby Mug of wholesome natural-ingredient A&W Root Beer with that true *draft* flavor you can't get in bottles or cans. Taste for yourself! It's FREE "Treat-A-Friend" Month at A&W. Bring coupon below for TWO FREE Root Beers at any A&W.

LOOK FOR THE BRIGHT ORANGE A&W!

Valuable franchises available in the U.S.A. and abroad. Write today A&W ROOT BEER CO.
922 BROADWAY • SANTA MONICA, CALIF.

BE A HERO... TAKE HOME A TREAT!

Watch their eyes light up when you bring in that big bright gallon of A&W. (Be sure and take home enough — and don't forget delicious A&W food!)

Copyright, 1965—A & W Root Beer Co.

ALL THE WORLD LOVES A & W!

2400 A & W's Coast-to-Coast and Overseas

FREE: TWO A & W ROOT BEERS!

Treat A Friend to a big, frosty A & W Root Beer — Coupon good for TWO FREE ROOT BEERS. Cash redemption value 1/20 of 1¢. Offer expires Labor Day, 1965. Offer void where prohibited by law.

TO A & W FRANCHISED OPERATOR — Mail Coupons to A & W Root Beer Co., 922 Broadway, Santa Monica, Calif. per instructions in the Redemption Plan sent to you.

ONE COUPON PER CUSTOMER

THOSE HAVENS FOR HAMBURGERS

Although many of the more elaborate drive-in restaurants in this country boasted a gigantic menu and a seemingly endless variety of foods, it was that American favorite known as the hamburger that rose to fame and fortune along the highways and byways. By the end of the 1960s, the stacked sandwich that could be held in one hand (and eaten while driving) dominated the "fast food" industry. With the approval of the automotive-powered customer, mom and apple pie were muscled out of the kitchen and the hamburger was christened the new icon of American convenience food.

The transformation didn't happen overnight. In the early part of the century, the hamburger was regarded with much suspicion by the general public. No one could really trust what restaurateurs with an eye toward profits were adding to their mixtures of ground meat. The "hamburger sandwich," as it was called in those days, occupied a lowly position on the menus at most greasy spoons across the country.

McDonald's has always been a haven for hamburgers. Back during the 1950s and 1960s, great-looking buildings of the type shown here were the standard issue for franchisees eager to get into the business. Today, the neon-clad dream is over; the environmental look is the new architectural style. *Howard Ande*

Individually owned and operated hamburger joints like Clown-Burger were at one time the majority along America's roadways. These days, most of the mom-and-pop operations that occupy space along once busy access corridors are closing down, making way for the corporate giants of fast food. This site was recently razed and the clown retired from service. *Michael Karl Witzel*

In 1916, the situation improved slightly after traveling fry cook Walter Anderson opened a tiny, three-stool lunch counter in Wichita, Kansas. The burger operation prided itself on quality ingredients and clean surroundings. After expanding to additional locations, the fledgling chain was named White Castle. It was an appropriate name: gleaming counters and floors gained customer confidence. Food preparation was in full view and everyone could see what was going into their lunches and how they were made. Unlike a few other places of ill-repute, burger patties that dropped to the floor were never picked up, dusted off, and slapped between a bun for the next unsuspecting customer to purchase.

As White Castles spread across the United States and similar operations opened for business, the regional bill of fare that typified the roadside food of the 1920s began a slow change. Gradually, the hamburger gained enthusiastic converts and worked its way into the hierarchy of foods. While the blue-plate special (with all the trimmings) remained the top entree at sit-down restaurants, operations that added on drive-in service began to augment their menus with new edibles. More motorists were asking for burgers, and it seemed a logical choice when one considered that it had to be eaten in cramped quarters. The combination of ground meat and a bun was an almost perfect design.

Keller's Hamburgers is a surviving drive-in that does a booming business in the city of Dallas, Texas. A prodigious amount of neon lighting has made this burger stop a local landmark and this colorful sign pulls in traffic. *Michael Karl Witzel*

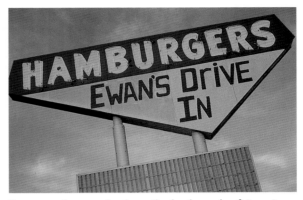

Every so often, a trip along the back roads of America reveals a drive-in or hamburger stand that isn't recognized by the consumer force-fed with continual television commercials and motion-picture cross-promotions. These are the places that one may get some of the best road food around. *Michael Karl Witzel*

In effect, the hamburger was a full-course meal without all the pomp and circumstance of the sit-down dinner. Unlike the loose meat sandwich, the circular hamburger stuck together (the addition of melted cheese helped). Because of its symmetrical properties, it could be attacked from any direction and rotated to suit one's fancy. Cheap chuck meat provided a protein source that was just as good as pot roast or steak, while simple toppings replaced the usual appetizers and dinner salad. Rather than being served in a table basket, buns could be integrated into the total package as well. Finally, hurried diners had a portable meal that contained fortifying items from all the major food groups.

Both drive-in proprietors and their patrons recognized the convenience, and by the 1940s the hamburger had gained a marked level of respectability along the roadways. While many a curb stand still served the various local specialties, the burger and its numerous iterations became a staple food. Many incarnations were tried, including double-meat stackups, cheeseburgers, chiliburgers, and more. Even so, a basic ingredient roster and preparation method emerged, defined by ground meat, bun, pickles, lettuce, tomato, and onions. Add-on condiments such as ketchup, mustard, and mayonnaise were slathered on in varying proportions, according to customer demand.

With the hot sandwich made of ground beef becoming the core of their business, the majority of drive-in restaurants in America became havens for hamburgers. Suddenly, frazzled commuters were ordering an inordinate number of these hot, circular sandwiches, forsaking the elaborate trout dinners, specially prepared soups, and other time-consuming entrees that they once gobbled down with glee. Life was getting faster and complicated all the way around, and as a result, affordable, portable, palatable food that could keep up with the times was poised for greatness.

A few drive-in operators with vision saw that the future of fast food was in the burger and decided that they would make it the basis for their entire business. Robert Wian was one of the first burgermeisters to gain fame, opening a small five-stool lunch counter in Glendale, California. In the winter of 1937, he was working the serving counter at Bob's Pantry when late-night customer Stewie Strange requested "something different." Wian

During the 1950s, the Harry Hines strip in Dallas, Texas, was the top place to cruise if you wanted to visit a drive-in. Although most of the classics are gone, the Prince of Hamburgers is one burger joint that remains. *Michael Karl Witzel*

cooked to please and the impromptu burger that sizzled off of the griddle became the signature two-story cheeseburger that earned him his reputation and promised to make his drive-ins a success.

At the time, six-year-old local boy Richard Woodruff was a regular customer at Bob's place. Woodruff was always looking for a free food hand-out, and occasionally, Wian let him sweep the floors or do some other odd job in exchange for a snack. Charmed by his droopy overalls, pudgy physique, and limitless appetite for grilled beef patties, Wian decided to call his new, multilevel sandwich the "Big Boy." Later, a local cartoonist sketched a rendition of the hungry street urchin on a napkin, and before the decade was done, the lad with tousled hair, red-and-white checked over-alls, and pot belly was a trademark adorning advertising signs, burger wrappers, and even the front facade of Wian's burger joint.

Aided by the memorable images of the Big Boy, news of the "double-deck" cheeseburger spread far and wide, and by the 1950s, Wian was franchising the tasty Big Boy sandwich and its

A few short decades ago, a successful hamburger seller need two things to remain profitable: a great roadside sign and good food. Today it's becoming extremely difficult for the individual operator to survive with only these two attributes. National advertising and name recognition have become an unfortunate requirement of the business. *Michael Karl Witzel*

In Sapulpa, Oklahoma, the Happy Burger that's doing business along old Highway 66 attempts to attract its patrons with a decidedly patriotic color scheme. Flamboyant paint is nothing new; burger joints have tried every trick in the book to get cars to stop and take a taste of their ground chuck sandwiches. *Jim Ross*

amusing trademark to restaurateurs in six states. Twenty years later, the portly kid was greeting customers nationwide. By that time, he had grown to become a larger-than-life statue sculpted of painted fiberglass, holding a deluxe cheeseburger platter high in the sky for all those passing to see. With a flavor unmatched, the Big Boy drive-ins went big time and carhops scurried about on the parking lots to meet the demand.

But Robert Wian wasn't the only one jumping on the hamburger bandwagon. Around the same time the Big Boy creation was gaining fans, Richard and Maurice McDonald were busy creating their own hamburger legacy. It all started when the pair were employed in Hollywood transporting movie sets. After taking notice of a local hot dog vendor who was doing a bang-up business, they quit their jobs to try their hand at the drive-in game. In 1937, they opened a small orange juice stand on Highway 66 and three years later, moved the octagonal unit to a better sales location in San Bernardino. For the next 11 years, they perfected their carhop operation until they could improve it no more.

In 1948 they decided that major changes were needed, and they temporarily closed the eatery. A few months later, they reopened with a more efficient operation. To their regular customers' great concern and shock, carhop service was eliminated, along with the once varied food items, the real silverware, the plates, and the choices! Along with fries, soda pop, and milk shakes, nothing but hamburgers would be served. Customers were required to make (and pick up) orders at walk-up windows. The McDonalds called this streamlined arrangement the "Speedy Service System."

Many of the classic, circular hamburger havens that copied the basic design of the ground beef sandwich were a great place to go inside and sit down—especially if the cold winter temperatures made the cheese on your cheeseburger turn as stiff as cardboard. *University of Louisville Ekstrom Library*

In 1940, the McDonald brothers sliced the "Airdrome" refreshment stand in two pieces and transported it to a new site located at 14th and E streets in San Bernardino, California. It was remodeled, and 20 carhops with satin uniforms were hired. On weekend nights, 125 cars touched fenders in the parking lot of this drive-in classic. *Richard McDonald collection*

After a brief drop in business, the concept really took off. In spite of the limited variations offered, hurried customers found the service to be impeccable and the food to be pretty good. Consumption of the hamburger sandwich that they had unanimously elected as the nation's favorite road food was now more convenient than ever! In exchange for the slightly more rigid business format, hurried diners could take advantage of economical pricing, lightning-quick service, and total convenience. Best of all, there was absolutely no tipping of carhops required!

Sometime later, salesman Ray Kroc discov-ered the magic of the setup, and in partnership with the McDonalds, took the idea of the fast food hamburger nationwide. By the time the 1960s were over, the mighty hamburger had spread across the land with little restraint, annihilating all of the homemade recipes and regional folk foods that were once eaten and adored by the cross-country traveler. Duplicated by an endless string of imitators, the "franchised" fast food hamburger became the body snatcher of the roadside restau-rant. In time, America would be oblivious to the qualities of drive-in dining that once made the act of consuming them so much fun in the first place.

Bob's Big Boy was quite fond of hamburgers and made it known that the grilled patty sandwich was a favorite of his. Clutching the proverbial double-deck burger in his hand, he graced the covers of menus for years. *Author collection*

Texas artist Randy Welborn has adopted the drive-in restaurant as a favorite motif for his evocative paintings of the American roadside. The circular Pig Stand drive-in shown in this scene is still a working operation that serves patrons in the city of Beaumont, Texas. *Randy Welborn*

Chapter Four

EVOLVING AMERICAN CURB FOOD

Although the hamburger sandwich has remained the paramount road food for more than four decades, many a motorist in search of quick sustenance along America's highways and byways has discovered that there's more to life than just a chunk of ground beef slapped between two halves of a bun. As the hamburger made its meteoric rise to the top of the junk food heap, a myriad of other portable foods that were easy to prepare, inexpensive to buy, and delightful to the taste buds found their own success along the roadways. Not surprisingly, they are still thriving today.

The big question: How did these other popular food entrees get a foothold in the business of roadside food? The answer is simple. Many got their start long before the automobile was accepted as a practical mode of transportation and as a direct result, were perfected by the time the internal-combustion engine permanently replaced horse power.

Susie's Drive-Thru is one of the finer fast food spots along Chicago's Montrose Avenue. Rather than serve up the same old fare, Philadelphia beefsteak sandwiches are a specialty here, offering motorists a chance to taste something other than the burger-and-fry combination. *Howard Ande*

Joliet, Illinois, is the place to find delicious Polish hot dogs when you're speeding in an automobile down Route 30. Naturally, it's also a good place to get a hamburger, as indicated by this neon-sign beauty captured at dusk. *Howard Ande*

When the motorcar became a common transit appliance, those food items that meshed with three very important criteria were poised for gastrointestinal greatness.

Among these important determining factors was the innate ability of a food item to be carried aboard an automobile without any additional fuss or bother. Warming pots, large serving tureens, or other such gadgets that were found in the finest kitchens of the day were not at all acceptable. Along with grandma's recipe book and well-stocked larder, there was no place for a cast-iron stove or pile of firewood in a motorcar—unless one was piloting a steam-powered vehicle.

In addition to portability, entrees that aspired to become the favored food of automobilists had to possess the unique attributes that allowed them to be consumed with neither knife nor fork. With that in mind, a successful car food placed minimal demand on the driver's manual dexterity. In light of unwieldy steering wheels (no hydraulic assist), starting cranks, dashboard levers, gearshifts, and other appendages that had to be controlled while in transit, it was a definite necessity that food taken on the road be easy to hold—and eat—with just one hand. If it could be consumed with a few large bites, all the better.

Since the upscale city restaurants, wayside inns, and well-appointed "tearooms" of the second and third decades were often rather expensive affairs, cost became a consideration in the formulation of successful roadway fare as well. The enthusiastic motorist of the day was required to spend a considerable amount of cash on the repair and upkeep of the motor vehicle. Consequently, all of the many food items that were to be considered as candidates for mass consumption along the nation's traffic corridors had to be economical and affordable.

So-called corner cafes such as this pastel apparition found in Grapevine, Texas, are a modern-day rarity. Only small towns and out of the way places that time has forgotten allow operations such as these to thrive—and serve food. *Michael Karl Witzel*

Even after meeting these varied guidelines, there remained one more need to be satisfied: Meals that were to be consumed by the cross-country adventurer had to be highly accessible. Motoring was still somewhat of a dirty hobby from 1910 until the early 1920s, and more often than not, it resulted in the participants getting soiled with dust, spattered with grease, or stained with gasoline. While they were wearing the dusters, goggles, and heavy gauntlets required by open roadster driving, enthusiasts were often reluctant to step into a "civilized" inn to request food and beverage. For the ardent automobilist, picnic baskets were all the rage.

Fortunately, there were a few operators of the time who perceived the growing demand for fast food that was served away from the home.

Using whatever culinary and marketing skills they had at hand, they rose to the challenge of feeding the growing motoring masses. At first, so-called greasy spoons and dining shacks known as "beaneries" sprang up—usually near the towns that hosted large factory facilities and a substantial amount of industrial workers who needed to take their lunch away from home.

Here, an assortment of basic food items (all using cheap ingredients requiring minimal culinary skill to prepare) like chili, beans, soups, and stews were served up. For the assembly-line worker who toiled across the street and needed a fast lunch, the arrangement was acceptable. But these low-buck joints were not geared to the free-wheeling motorist. All of these facilities demanded that customers arriving by foot or car sit down at a table or counter.

The roadside diners, cafes, restaurants, and drive-ins of America were once great places to hear your favorite singing group. When inside seating was available, booths were almost always outfitted with jukebox controllers that provided three plays for a quarter. Often, the selection of songs available was as eclectic as the many choices on the food menu. *Michael Karl Witzel*

Serving beer used to be a really big deal at drive-in restaurants like Keller's in Dallas. In order to get around local ordinances that made certain areas "dry," operators often moved outside the city limits so that the frothy brew could still flow freely to the customers who wanted it. *Michael Karl Witzel*

Leslie's Chicken Shack was a Waco, Texas, restaurant that employed curb service for a number of years. Serving "California-style" fried chicken, the eatery opened operations in nearby cities and continued to maintain this unit until recent years. Now, the famous chicken sign and building are up for sale. Another unique roadside food has gone extinct. *Michael Karl Witzel*

While Harlan Sanders, the "Colonel," is best known for his secret Kentucky Fried Chicken recipe, there remains a gaggle of drive-ins throughout the states that serve up fowl with equal flavor. This Chicken 'n Dip serves both ice cream and bird and is located along Route 72 in Hampshire, Illinois. *Howard Ande*

The Iceberg is a small-town Kansas favorite that uses a Buffalo Burger specialty to attract its clientele. The citizens of Fredonia (and those just passing through) visit it often for the unusual bill of fare served there. *Michael Karl Witzel*

As more and more of these lunch spots opened for business, problems with sanitation, hygiene, and food storage began to cast a dark shadow on the low-cost dining routine. Seventy-five years ago, the healthful tenets of food service that are followed today were seldom adhered to at many of these one-man stands. With only crude methods established for food preservation and a total lack of refrigeration, many of these dining establishments gained a somewhat dubious reputation regarding the quality (and sometimes taste) of their food.

Because many of these unsophisticated restaurants were mom-and-pop enterprises with little capital at their disposal, the personnel that prepared the food often lacked training in the art of cookery. Irrespective of cold sandwiches (not as

popular then as they are today), people demanded hot things to fill up their bellies. As a result, the select recipes that could be created with great haste were often centered around some sort of integrated food item that was made from a mixture of ingredients. Ground beef or pork—in the form of circular patties, meatballs, or cased sausage—was seen as the perfect fodder for the untrained chef.

Unfortunately, these ingredient caused a few problems with quality. Many turn-of-the-century proprietors took advantage of the opportunity and adulterated their products with additional fat, filler, and inferior additives. Unlike a bowl of beef stew or beans that could be carefully examined as each of the individual spoonfuls was lifted toward the mouth, bites that were taken directly from encapsulated food items left scant opportunity for

At the Hot Shoppes, barbecue made of ham and pork shoulders was the main bill of fare. With hot sauce and relish, it sold for 30 cents during the 1950s. The famous milkshake was next; at 25 cents it got the motorists' attention more often than root beer. As thick as a frappe, it was modeled after the delicious shakes made at the Brigham Drugstore in Salt Lake City where young J. Willard Marriott attended college. Although gas rationing blacked out curbing for several years during the war, these staple entrees never lost favor with the public. By the 1950s, 2,000,000 barbecues and 2,300,000 milkshakes were sold per year. The question was, who came up with the trademark orange roof first, Mr. Marriott or Mr. Johnson? *Bob Sigmon collection*

a detailed examination. It was ironic indeed that the very attributes that made hamburgers and hot dogs a perfect choice for car food also relegated them to a questionable status. Until improvements were instituted throughout the industry as a whole, the convenience of dining along the highways and byways were overshadowed by cases of stomach cramps and intestinal distress.

Those changes were quick in coming, as more and more automobiles hit the streets, the unscrupulous minority of food concessionaires were effectively forced to change, or exit the business. By the early 1920s respectable drive-in restaurants like the Texas Pig Stands began to appear on the American roadside with greater frequency. Gradually, they altered the course of roadside dining and began repairing the damage done by their unsavory predecessors. With an eye toward expanding operations nationwide, these roadside upstarts initiated a trend that diverged from the saucy variety of sit-down foods that the majority of lunch counters were serving and opted for meals that could be eaten with more ease.

By maintaining simple menus and featuring one food item as their specialty, they were better able to get a handle on the variables that caused

When Roy Allen and Frank Wright teamed up to sell root beer in the early 1920s, they decided to combine their initials and form a new company name. A & W was the result, incorporated into a graphic symbol featuring the now-familiar "pointing arrow." Since the day the first frosty mug was sold in 1919, a mug of A & W Root Beer could be purchased for just one nickel. By the time World War II broke out, the "5-cent" circle was as much a symbol of A & W as the orange-and-black logo divided with an arrow. (Notice the direction.) *A & W Restaurants Inc.*

Chicken in the Rough was invented in 1936 by Mr. and Mrs. Beverly Osborne. They ran a small Oklahoma City drive-in and franchised their tasty poultry dish to operators nationwide. Consisting of one-half a golden brown chicken served with shoestring potatoes, hot buttered biscuits, and a jug 'o honey, it was a meal served without silverware. In 1958, a Chicken in the Rough platter could be purchased for $1.40 Tuesday through Sunday. It was reduced to $1.00 on Monday, which was family night. *Jim Ross*

foods to become tainted in the first place and at the same time control quality. Barbecued "pig sandwiches" were a great choice, for example, since the meat used to make them was maintained at the proper temperature in a large smoker. At the same time, bread was delivered daily and soft drinks were served out of sealed bottles. The popular press took notice of the many drive-on success stories, and as the incidents of ptomaine poisoning declined, the notion of the specialized food stand gained increased acceptance with America's motorists.

Curiously enough, the same convenience foods that were once regarded with suspicion rose to the top of the table in roadside fare. By way of the drive-in restaurant, selections that were affordable and could be prepared quickly became the standards for a nation that was inebriated by the limitless freedoms afforded by the automobile. The ability to wolf a hot dog, chomp on a hamburger, or gnaw on a basket of fried chicken was appreciated by the open roadster crowd that had little time to waste with the excessive formality of dining in an indoor eatery.

Chapter Five

DRIVE-IN DIVERSITY

*T*oday, the architecture of roadside food operations has reached a defined level of banality. It all started during the 1970s when operators decided that it was no longer acceptable to look just like a plain old burger joint. Exterior panels of porcelain enamel were removed, neon signs were taken down, and unusual design qualities muted. Suddenly, hamburger huts that catered to car customers were required to "blend in" with the environment.

All at once, the commercial design motifs that had served the public for so many years were deemed outmoded. In the rush to upgrade and constantly create a new and improved image to their expanding customer base, the brain trust at the helm of America's restaurant concerns decided that drastic changes were in order. The postwar configurations that had brought them so

The city of Great Bend, Kansas, is the place to be driving if you are searching for a genuine example of programmatic drive-in architecture. Today, a surviving Twistee Treat ice cream stand illuminates the night and does a brisk business from within a colorful building that's shaped like an ice cream cone and hot fudge sundae rolled into one.
Howard Ande

49

much success were to be re-engineered and redirected to project a softer, more friendly attitude to the general public.

This kinder and gentler architectural slant resulted in the hasty demolition and remodeling of some of our nation's best examples of drive-in design. With little protest mounted by customers, the gleaming boxes of bright porcelain enamel were hastily resurfaced with brick. Flat roofs and dramatic advertising pylons that once captured imaginations were now considered passé and the simple mansard roof emerged as the new favorite. The chrome-plated look that was so indicative of yesteryear was dramatically dropped. Natural woods of every imaginable type became the universal replacement.

Out on the restaurant parking lots, it was no longer enough to provide a clear, flat, accessible pad for the arriving automobiles. Landscaping was taken to the extreme, characterized by a maze of predetermined driving lanes and decorative greenery that was integrated into the total restaurant package. Now, the burger-and-fry bunch were to imagine that they were visiting just another comfortable home in the neighborhood, complete with all the trappings of suburbia. At the same time, a separate playground area for the children became a prerequisite to convenience dining as well. Meticulously maintained lawns (emulating a golfing green), fine trimmed shrubbery, and decorative flowers of the finest variety completed the bucolic fast food illusion of the new age.

Seventy-five years ago, the entrepreneurs who were eager to reap the benefits of the automobile were not so concerned with finding an architectural aesthetic that appealed to the universal mind. In those early days of the drive-in, most wayside dining operations that served the newfan-

Elaborate architectural beauties like the Uptown Cafe in Branson, Missouri, routinely use the building styles that made the drive-in restaurant such a major attraction. Glass block, neon tubing, and polished metal are some of the elements that recall the crazy days of carhops and curb service. *Howard Ande*

The Triple XXX Corporation of Houston, Texas, sold a lot of root beer during the 1930s and 1940s. Outlets were generally located in the Western half of the states, with a profusion of locations stretching from California and Washington. Key to the operation were buildings incorporating a giant root beer barrel. Many featured larger-than-life renditions—while some (such as this Washington State version) added a false front facade shaped like a barrel. As the slogan said: Triple XXX "Makes Thirst a Joy." *Michael Karl Witzel*

gled horseless machines were tentative constructions at best. The motorcar was still an unproven machine, and the outlay of large cash investments in hope of snaring this new market was risky. Unadorned shacks and shanties built with available lumber and "found" materials ruled the day.

There was one practical reason for this rather naive design philosophy: At the turn of the century,

In Globe, Arizona, a teepee-shaped pizza restaurant ekes out an existence. While more than 75 years ago the idea of a building as sign was all the rage, today it's not so appealing. The motoring public has become jaded to out-of-scale structure, and the elaborate signboard has assumed the responsibility of highway marketing. *Michael Karl Witzel*

The Mammoth Orange stand drive-in operation was established by Peggy Doler, now retired and living at Garden Grove, California. During the 1940s, she began the operation in her front yard to take advantage of the business rolling by on Highway 99. With the arrival of the new freeway, she was bypassed. As her lifeline was redesignated Chowchilla Boulevard, business dropped. Doler moved the hut to its present site in 1953, and five years ago Jim and Doris Stiggins purchased it. Today, truck drivers still ask about Doler and reminisce about the days when the spring opening meant free food for all of the truckers. Some customers say they have been stopping there for more than 25 years. Baby-boomers recall wheeling in for a chilled flagon of juice with their parents! Taste buds never forget. *Glen Icanberry*

roadside operators were rather inexperienced when it came to appearances. With the automotive industry still in its infancy, there were no franchising guidelines to show them what to do, no books written on the subject outlining the fine points of roadside salesmanship, and no great business leaders to serve as mentors. As a result, those who wanted to get into the fledgling drive-in restaurant business had to make things up as they went along—discovering the hard way what would and wouldn't work.

The passage of years proved that the automobile was definitely here to stay. When the car replaced the horse for good, businessmen began investing ever larger sums of money into their roadside endeavors. By the end of the 1920s, the economic prosperity and giddiness of the times was reflected along the nation's roadsides. "Programmatic" architecture began to appear with more and more frequency, characterized by roadside structures that broke all the rules of conventional building design. Whimsical themes, crazy concepts, and other personal flights of fantasy were made into three-dimensional structures to sell food.

With eye-appeal that promised to draw in car customers like bees to honey, the drive-in restau-

California architect Wayne McAllister's simple circular structure (popularized by restaurateurs operating during the 1930s) has even influenced styles across the Atlantic. Copenhagen, Denmark, is home to the Cafe Around the World, a drive-in throwback that features both indoor and patio dining. The neon advertises a popular Danish beverage, Tuborg beer. *Michael Karl Witzel*

rant was ideally suited to the flamboyant style of programmatic architecture. Limited only by the imagination of those who constructed them, buildings of every conceivable type appeared along the highways: To drive home the fact that sudsy beverages were their mainstay drink, operations such as A & W Root Beer erected fantastic stands that resembled gigantic root beer barrels! At the same time, they experimented with even more mesmerizing, out-of-proportion constructions such as oversized Indian chiefs and scaled-down lighthouses.

Captivated by the dramatic effect of these attention-getting designs, imitators such as the famed Triple XXX root beer stand (located on Waco's busy traffic circle) added oversized kegs as an adjunct to their structure. Animals and other familiar creatures became favorite drive-in motifs as well. By the end of the 1920s, it wasn't unusual to snack on a hamburger at a diner shaped like a giant toad or even a pig. Typically, the program that defined the shape of a food or drink building was indicative of the food specialty served there.

In spite of its strong visual appeal, the "building as sign" motif eventually faded, and by the 1930s the business of the drive-in restaurant was well established in America. With rising competition and the need to turn over customers with increased speed, more serious styles of restaurant architecture were being considered. Unlike those early pioneers who simply threw up a wooden shack and a few hand-painted signs (and opened for business the very next day), modern restaurateurs began looking in earnest toward the professional architects for the answers they needed.

With its expansive boulevards and weather

Highway 54 used to be the main route entering Kansas from the east. Before the freeways, it was the conduit to Wichita—thick with roadside businesses eager to cater to the motorist. Today, this strip of asphalt known as "The Yellow Brick Road" is a living museum of roadside attractions. *Michael Karl Witzel*

conducive to year-round drive-in dining, southern California led the way with designs that were completely dedicated to car food service. Architects implemented a new-and-improved vision for car commerce, and by World War II they had established the "circular" style drive-in as the reigning standard. Appropriately, the balanced-wheel motif proved to be the perfect configuration to sell drive-in car dining to the rest of the nation—incorporating into one flamboyant package all of the attention-getting elements needed to pull paying customers off the streets.

Before too long, nearly every major street corner in Los Angeles played host to a circular drive-in restaurant. From the vantage point of the busy boulevard, these dramatic sculptures of chrome, glass, and stainless steel were truly a sight to behold. Free-floating roof overhangs appeared to defy the laws of gravity as extravagant center spires shot skyward, heavily bedecked with miles of neon tubing. The unified look that was offered by this circular arrangement transformed what was at one time a mundane roadside concession stand into an attractive space station for cars.

Still, there was more substance to the circular design than just heightened visual aesthetics. Ever since the days of the first curbside dining stand, problems with traffic flow and serving logistics had been major concerns. A round design was perfectly suited for roadside dining; with unprece-

When man's race for space began influencing the way designers perceived architecture and automobiles, subtle reminders of that hopeful futurism began appearing as embellishments. While motorcars grew tailfins and rocket-like protrusions, roadside stands such as this Virginia U. S. 1 Dairy Bar added modernistic structural elements. *Michael Karl Witzel*

dented ease, drivers could access the parking lot, cruise around the building's periphery, and quickly shoot into a parking spot that suited their fancy (the "inner circle" became the place to be). Because every dining space was nearly an equal distance from the kitchen, the carhops were better able to maximize their movements.

With the acceptable range of drive-in architecture widened by a few architects with great vision, restaurant operators located in other regions of the country were liberated to interpret their very own way-out drive-in restaurant dreams. And they proceeded to do exactly that, incorporating dramatically stacked layers into their overall building design. By the conclusion of the 1930s, electrical illumination schemes, glass blocks, porcelain enameled panels, streamlining effects, flashing neon, polished metal alloys, stucco exteriors, protective overhead car canopies, elaborate tilework, concrete parking lots, automated serving gadgets, and much, much more had been tried.

Without a doubt, the drive-in restaurant had come a long way and would remain the architectural gem of the American roadside for decades. Taking prominence over the hotel, motel, and gasoline service station, the buildings that allowed patrons to eat in their cars had become an inseparable part of pop culture.

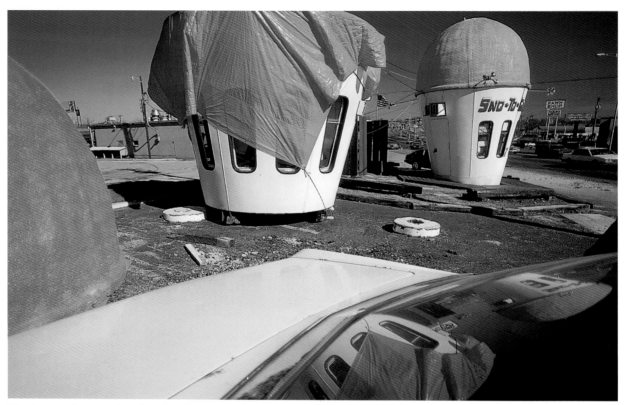

Playing on the visual and verbal language that influences through unconscious association, the imagery of the sno-cone stand captures our attention. "Hey, look at me" is their flashy call to travelers on the open road. In their attempt to break through the insistent cacophony of competing roadside images, they easily succeed. We are lured from the speeding fast track to enjoy the tasty wares of this unique curbside attraction. As synthesis of sign and building, their mimetic form of architectural design reeks of pure kitsch. *Michael Karl Witzel*

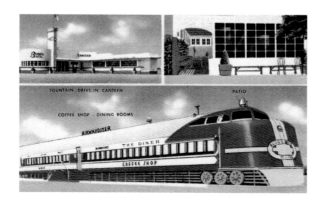

The Streamliner Diner Drive-In was one of those cool roadside eateries of yesteryear that made you want to stop just because it looked intriguing. Not only was it a restaurant and a diner, but it also featured carhop car service right from an adjoining drive-in. The train has always been a popular motif for eateries and even today, one may find theme restaurants housed in retired boxcars, old engines, and more. *Author collection*

Seattle's Igloo Drive-In was a well-known attraction during the 1940s. Inside seating for 70 customers made the dual-domed wonder an attractive destination for those who liked to eat in an enclosed dining room. Under the stars, wintertime carhops wore ski-togs from Nordstrom with high white boots and in summertime, short skirts. Owners Ralph Grossman and Ernie Hughes recruited most of the good-looking girls from the local theaters to work there as carhops. *CoolStock/Ralph Grossman photo*

Chapter Six

CRUISING THE MAIN FOR CURB SERVICE

More than any other roadside business in American history, the drive-in restaurant occupied a unique position in the continually evolving culture of the automobile. From the early days of the 1920s until its rapid fall from grace during the 1960s, the curb-service eatery held the distinction of being *the* universal meeting place for this nation's youth.

While the prewar decades saw their share of revelers whooping it up at the nation's drive-in diners, it was the generation that came of age during the 1950s that latched onto the idea of in-car dining with a fervor like no other. And why not? The promises of postwar prosperity were directly reflected in the drive-ins of America. Colorful bands of neon advertised that "happy days are here again," and natty carhops zipping about on roller skates conveyed a reassuring sense of order.

Under the canopy is the place you want to be if you want to be part of the main action that's going on at Keller's Dallas, Texas, drive-in restaurant. Just like during the 1950s, cars jockey for position in the service lanes as the day turns to dusk. By nightfall, a variety of hot rods and street machines may sometimes be seen there. *Michael Karl Witzel*

Along Route 66 in Seligman, Arizona, Juan Delgadillo entertains (and feeds) customers at the Sno-Cap Drive-In. Over the last few years, it's become a favorite drive-in destination for people who are cruising along the "Mother Road" and wish to stop and grab a taste of yesterday. *Howard Ande*

Always first to catch (or create) a new trend, American youth quickly adopted the drive-in as their personal stomping grounds and welcomed it into their clique with open arms. When they weren't sleeping, attending school, studying, or playing sports, the parking lot at the local drive-in restaurant was their favorite haunt. There, young adults finding their own way in an adult's world could meet with old friends, find new ones, seek romance, and pursue all of the many social activities that define teenhood.

At the same time, they could escape the confines of the suburban dining room and get a break from the same old meatloaf and mashed potato dinners. By the mid-1950s, American drive-in restaurants were bypassing many of the elaborate entrees of their restaurant brethren in favor of the "junk foods" that kids adored. The fact that burgers and fries were greasy and not particularly healthful for a growing body was inconsequential. Soda pop overloaded with sugar, you say? It simply wasn't an issue in those days. Drive-in food was fun to eat and virtually impossible to get at home.

But there were even more important aspects of the drive-in restaurant that ensured its success as a central gathering place. Around the circular hub of the service wheel, teenagers could congregate with others who believed in the same things that they believed in and who spoke the same language. Almost every high school had a favorite

The Healthcamp Drive-In is a modern-day eatery that's a favorite with cruisers in Waco, Texas. An outfit that calls itself "The Heart of Texas Street Machines" has adopted the eatery on the circle as their favorite destination. *Michael Karl Witzel*

drive-in that they adopted as their own. On any given moment of any given day, somebody from school could be found parked there. After school, the local drive-in appeared to be a magnet for cars. From far and wide, participants from all social and cultural backgrounds were drawn to the brightly lit canopies for a few enjoyable hours of food, good times, and fun.

And great fun it was, especially if the pastime known as "cruising" was involved. Enjoyed even before the car existed, albeit in a somewhat more refined form, cruising evolved to become a purely American activity that involved a central strip of in-town roadway and an automobile (a full tank of gas

definitely helped the proceedings). To participate, teens piled into a car, preferably a cool-looking one, and drove around to the many hot spots situated nearby. Usually a major part of this localized road trip involved a slow crawl down the main street corridor. The object was to see and be seen.

As the night's cruising progressed, highly favored routes of travel, or "loops" as they were called, were devised by the participants. Along the way, the many drive-in restaurants that opened for business during the 1950s provided the perfect place to stop and regroup. Because they were the most visible venues at which to be seen and the most likely spots to elicit reactions from others,

When vehicles are parked at a cruise night or other car show, their owners often mount car service trays—complete with wax foods and other drink replicas—to bring about the atmosphere of the American drive-in restaurant. Cruising and curb service are forever linked in the mind of the cruiser. *Kent Bash*

drive-ins became an important part of "shooting the loop." While cruising the main street strip was fun, nothing compared to the many activities that were going on down at the neighborhood Mel's, Duke's, or Cappy's.

Because of the conspicuous theater that they provided, the goal of cruising was to visit as many drive-in restaurants as possible in a single night. From dusk until the wee hours of the morning (when all the joints closed), cruisers flitted from one drive-in to the next and rode the same track into the dawn until the route was burned into memory like a phonograph needle stuck in the groove of a record.

Cruising was more than just the opportunity to make appearances, however. A big part of the pastime was the serious quest to meet members of the opposite sex. In a never-ending quest for love (or even just a glance) in the fast lane, guys in souped-up hot rods revved their engines at the lights and preened with all the bravado that they could muster, just to attract some girl's attention. At the same time, gals cruised the streets in their own cars (or mom and dad's) to be pursued. The final moves in this age-old game of pursuit and courtship were played out beneath the flickering marquee of the American drive-in restaurant.

In California, the "cruise night" has become a regular part of the weekend. Every Saturday night, the owners of classic hot rods and custom cars take their prized possessions out of the garage and drive it to their nearest drive-in. Until late night, they eat, talk, and reminisce about the "good old days." *Kent Bash*

While boy-meets-girl dramas played out around them, other drive-in revelers pursued more dangerous games under the auspices of outdoor automotive dining. Most visible among these secondary activities was the obsessive desire of young men to compete with one another. When automobiles were involved, that competition went far beyond mere mechanical prowess. Illegal, high-speed competition out on the streets, one car against another, was the best way to prove one's manhood to the peer group.

Most of the illicit drag racing heats run during the 1950s were arranged at the drive-in restaurant, since that's where the hottest cars and their owners were often parked. Hot rods and cool custom cars were a regular fixture at the drive-in, with the many carhop lanes occupied by gearheads peering intently under their hoods and polishing chrome or tweaking some high-performance goodie. When an interloper with a tough-looking ride pulled in to make a challenge, arrangements were made to meet at a clandestine drag strip usually on the outskirts of town. Like two rabid dogs straining at their leashes, both cars roared out of the drive-in eatery to meet their fate. Only one would claim the victory.

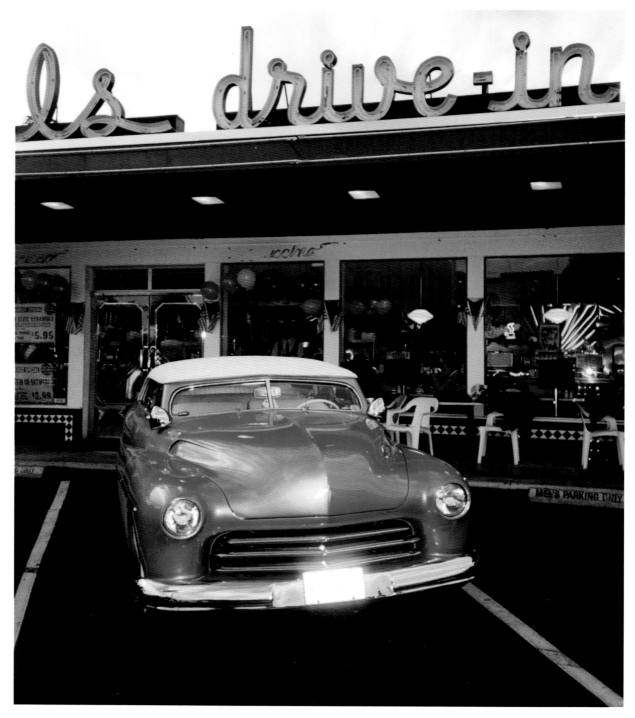

Oscars Drive-In provided the cruisers of the 1950s and the 1960s an appropriate venue to hang out and display their hot rods. Unfortunately, the popular dining spot was eliminated in the name of progress and the cruising fun that was had there remains only as a recollection. *Andy Southard, Jr.*

Hunger, a station wagon full of screaming kids, and a busy mother meant only one thing during the 1950s and 1960s: a trip to the local Steak n' Shake! With window trays stacked to the hilt with Tru-Flavor milkshakes and all sorts of tasty food delights (with chili), it was a time that any hungry child would remember for the rest of his or her life. *CoolStock/photo courtesy Steak n' Shake, Inc., sign Shellee Graham*

The Mercury is one of the standards of the custom car set, as is the Geary location of Mel's Drive-In. This location does a thriving business in the heart of San Francisco and is a favorite place for the locals to dine. It features inside seating that exudes a 1950s-type of nostalgia. *Michael Karl Witzel*

Dallas, Texas, was the origination point of the drive-in eatery and, consequently, had more than its share of curb service restaurants. The Circle Drive-In was a favorite among the teenage crowd of the 1950s, and it became a popular hangout for local students eager to socialize and get a meal. *Dallas Public Library*

On a flat quarter-mile stretch of road, two cars lined up and waited for the signal to hit the gas. If the coast was clear and there was no indication that the cops were cruising nearby, off they went, taking up both lanes of the highway until the faster vehicle pulled ahead and the defeated one broke off. With the winner decided, both racers and the thrill-seeking crowd returned to the drive-in where the winner enjoyed bragging rights for the night. In the movies, it was at this time that the champion met up with the girl of his dreams and they both drove off to meet the dawn of a new day.

Paradoxically, it was the ever-widening popularity of the drive-in restaurant as a cruising destination that contributed to its ultimate downfall and demise. With so many overstimulated teenagers hanging out there around the clock, trouble erupted all too often. On occasion, groups (gangs) from competing schools or car clubs traveled to drive-ins in other towns to stir up trouble. Even among the home team, the friction began to build by the early 1960s: Greasers clad in black leather jackets and ducktail hairdos clashed with socialites in lettermen sweaters. By that time it was more than just Chevy versus Ford; it was the cool against the uncool, black against white, hippie against jock.

With kids in their cars hanging out all night, the drive-in became a no-man's land for the family crowd. In neighborhoods that bordered the popular spots, people complained about the never-ending din of fights, cars burning rubber, and loud rock and roll music. Before too long, the authorities enacted ordinances to restrict the teenagers' movements. Drive-in restaurants began instituting all kinds of rules and regulations, and the entry of cars was closely scrutinized. Now, going to a curb spot to hang out or "loiter" was against the law, as was the repetitious act of cruising up and down the strip. By the 1970s, the freewheeling atmosphere of fun and frolic that was once enjoyed at the drive-in restaurant had diminished considerably. The heyday of cruising the Main for curb service had ended.

By the 1960s, the cruisers visiting the drive-ins were seeing a change in the way carhops looked and dressed. The styles of the era dominated their appearance and it wasn't long before all of the curb girls were decked out in mod hairdos and bell-bottom pants. *Dallas Public Library*

ELLIS DRIVE-IN
Theatre and Restaurant
PHONE VI. 2-2784
ROUTE 50
Clarksburg, W. Va.

CURB SERVICE

Close Cover 61 For Safety

THE RIO CLUB
Fine Music Entertainment Casino
Pocatello's NIGHT SPOT
Pocatello, Idaho
ONE MILE WEST OF POCATELLO
CLOSE COVER BEFORE STRIKING MATCH
THE DIAMO...

Fred & Kelly's RESTAURANT
GOOD FOOD
ALWAYS OPEN
AIR-CONDITIONED
Made in U.S.A.
THE DIAMOND MATCH CO. N.Y.C.

Varsity Waitress Is -
HAPPY TO SERVE YOU
LOCATION
GUEST CHECK NO.
12701

BIG CHAMP

X-S...

SENSING the swing of Swank
Wilshire ...levard, we selected
corner (at... for our second
There ...nd that
standa...count...
Then ...
pat...
R...

THE WILSHIRE

...in the hea...
...ahuenga, ...
...ber Thre...
...er sup...stupen-
...ine dining room; lunch
...onvenient counter or
...here you can
And for cocktails,
...here the clan
...ssal!

BURBANK" ... Victory and Olive Boulevards,

ENCORE OF YOU...

Some Home

Chapter Seven

ARTIFACTS OF THE DRIVE-IN ERA

*T*oday, most of the drive-in restaurants that populated the American roadside during the 1940s, 1950s, 1960s, and even 1970s are gone. On occasion, modern-day motorists may stumble upon decaying remnants of a defunct eatery or a broken sign that once attracted customers with its friendly glow—but these discoveries have quietly become the exception rather than the rule.

With every passing year, the physical evidence of the classic American drive-in restaurant disappears from the roadside scene. With little regard for posterity, the bulk of these former car havens, ice cream shacks, and burger huts are being slated for demolition. To replace them, an endless parade of bigger parking lots, strip malls, and franchised fast-food operations are being planned. All too quickly, the once grand burger huts are fading from memory.

Matchbooks were at one time given away free by the majority of businesses in America. Drive-in restaurants, diners, dry cleaners, and other businesses passed them out to anyone and everyone. Today, they are hoarded by many collectors and join similar articles as reminders of long-defunct operations that once made up the sides of the American streetscape.
Michael Karl Witzel

"Legs," one of Stanley Burke's shapeliest carhop girls, adorned streetside signs, menus, and dinner platters. When Stan's was at the height of its popularity during the 1950s, food was still served with real silverware and plates. For the carhop, carrying two fully loaded trays was formidable. *Michael Karl Witzel*

Fortunately, the foresight of a small minority has allowed for the preservation of a portion of this tasty legacy for future generations. While most of the original architecture typical of the classic styles of decades past has been replaced, a few select artifacts remain that readily conjure up those days of triple-thick milk shakes, double-deck hamburgers, and roller-skating carhops. For curb service aficionados, that's good news.

Still, a few major drive-in touchstones have been saved. Many rare operations that are still in business (such as the Texas Pig Stands) are choosing to restore their big neon signs rather than replace them. Neon that does make it to the junk heap is often snapped up by collectors with the cash to transport it home. But more often than not, the kind of signs that end up in the suburban den are of the simple one- or two-word variety like "Eat" or "Food." The larger, more elaborate marquees end up in museums such as Henry Ford's Greenfield Village or as the centerpiece for a 1950s-retro theme restaurant chain.

For the average drive-in memorabilia collector, a more practical direction is afforded by the various elements that at one time made the drive-in work. Everyday dining utensils such as platters, silverware, drinking glasses, milk shake tumblers, coffee cups, root beer mugs, and window serving trays were at one time an integral part of feeding people in their cars. Best of all, many of these

If one could travel back to an A & W operation of the 1940s, this scene is probably what they would see. These days, heavy glass mugs of the type shown, along with the serving tray that cradled them, have become sought after artifacts to memorialize those drive-in days. *Michael Karl Witzel*

Some collectibles cross over into other categories and as a result are much higher in price and much harder to find for the aficionado of drive-in items. All major soda pop manufacturers produced internally illuminated display units similar to this example, and today they pull down top dollar among those who appreciate vintage ad art. *Michael Karl Witzel*

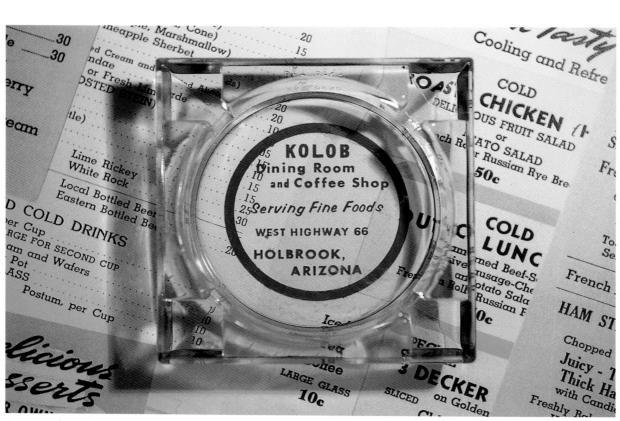

As a simple and easy to obtain artifact, the glass ashtray imprinted with the name and address of a drive-in or other roadside eatery is readily obtained. These holdovers from the days of unlimited smoking in public may be found at flea markets and garage sales nationwide. *Michael Karl Witzel*

practical dining goods were imprinted, engraved, fired, or embossed with logos and trademarks. Since a fair number of car customers sped off with these items to keep as mementos, they are now available to connoisseurs in limited quantities.

Among these "artifacts," the heavy glass serving mugs such as those used by refreshment operations like the A & W Root Beer stands have become immensely desirable. Since the first days of the drive-in restaurants an endless variety of root beer brands were introduced, and it seemed that most all of them had mugs that touted their particular brew. Naturally, the temptation to keep these unique tumblers was overwhelming for

many who liked to drink the sweet, frothy beverage. Many ended up being shoved under the seat and taken home for personal use.

Some car patrons with a bit more nerve decided that the actual food plates, often imprinted with the whimsical depictions of carhops and other such drive-in scenes, would also make neat keepsakes. On one trip, it was quite easy for a plate to be surreptitiously stashed in the glove compartment (40 years ago, these were spacious car cupboards) and on another trip, a coffee cup could be stowed along with the saucer. Over the years, customers who had an inclination toward kleptomania could slowly

Linen postcards are like miniature time capsules. They provide today's viewer (and collector) with a rare opportunity to see what a real drive-in restaurant looked like and the colorful imagery to make the imagination wonder. *Michael Karl Witzel*

acquire an entire set of dishes from their favorite curb service eatery! Is there any wonder why today's fast food restaurants have switched to paper and plastic?

Another important part of the classic drive-in operation was the menu. While a few of them had signboards that listed all of the fare that was available, most of the major operations in California and the Southwest used printed menus to make their entrees known. In the days long before the current limited repertoire of car food outlets

became the norm, elaborate guides showcased the amazing variety of foods that could be served in the front seat of your car. Because each outfit had its own specialty and style of service, these colorful menus became a direct reflection of an eatery's personality.

For today's collector, the drive-in menus that somehow managed to survive and repel the multiple decades of spilled coffee, dripping hot fudge, hamburger grease, and gallons of soft drinks are an excellent way to recall those heady times. After

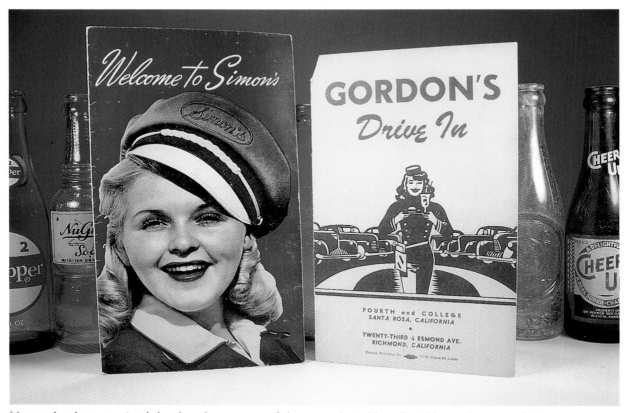

Menus that have survived the decades are one of the most desirable collectibles in the area of drive-in restaurants. Their appeal is varied, providing modern-day scholars a chance to see the changing trends in print advertising and roadside food service. With their splashy colors, funky design, and low prices, they are both a kick to look at and can make a great addition to anyone's collection of carhop memorabilia. *Michael Karl Witzel*

all, the menus that were designed some 40 years ago were bona fide works of art—highly graphic and colorful examples of the day's advertising art. At the same time, they provided a visible showcase for the fashions and attitudes of the carhop waitress. More than any other motif, the combination of a curb gal's smiling face and shapely silhouette was shamelessly exploited on the front cover of nearly every drive-in menu.

While menus were not intended to be carried away, the lion's share of American drive-ins invested capital in printed promotional giveaways of various types. Classified as "ephemera" in the present-day

jargon of collectors, a substantial resource of these fragile handouts have survived the passing decades. In much the same way as the menus do, these advertising artifacts provide the enthusiast with an unobstructed glimpse into the roadside past. Whether these portray "the good old days" or not is up to the individual collector to decide.

In this category of cool stuff culled from the drive-ins, postcards of the penny variety are one of the most hoarded items. At one time in history, they were a major portion of the roadside business ad budget. Across all of the 50 states, it seemed that every wayside snack stand and curb-

"The Burbank" Roberts Brothers drive-in located on Victory and Olive boulevards featured a twisted flash of what appeared to be electricity at the top of its pylon. With circular design, modern equipment, and air-conditioning for inside diners, it was a mecca for motorists in the San Fernando Valley. Today, only menus remain as evidence of the chain. *Author collection*

The fluffy-top soft-serve ice cream cone is a frequent motif employed to attract the commuter from his or her route. Everyone loves ice cream. Too bad this sign is all that remains of a former roadside business. *Michael Karl Witzel*

side refreshment stop (large or small) had some sort of imprinted postcard that they handed out. Upon them, satisfied customers touted the specialties that they had consumed and mailed the comments to their friends and family relations. It was a win-win situation for all: Restaurants would get some great publicity and the patron a freebie.

Currently, the so-called "linen" variety of postcards that were produced in great numbers by the Curt Teich Company of Chicago are the most desirable. An enterprising immigrant from Germany, Teich got the notion to make postcards of all types of roadside businesses as far back as 1898. He traveled the country in search of prospects and as a result, myriad roadside operations—drive-ins included—were immortalized in the aspect ratio of 3 1/2x5 inches. Today, the distinctive texture of the crosshatched cardstock that was once used for Teich's tiny billboards is a big part of their vintage appeal, as are the fanciful col-

The jukebox was a mechanical device designed to organize and play grooved recording discs for the public. A customer dropped a coin into a slot and selected a song from a list. Push buttons were pressed and a mechanical arm placed the "45" on a turntable platter. A moveable arm housing a "needle" rode the tiny tracks in the disc and transferred minute variations in the vinyl into electrical impulses. An amplifier fed these pulses into a loudspeaker. Before radio was big, musicians relied on jukebox play to gain notoriety. Here, a remote-controlled unit is seen at a California Mel's restaurant. *Michael Karl Witzel*

In 1949, Leo S. Maranz followed the success of Dairy Queen and designed his own automatic ice cream freezer. Unfortunately, he learned that it was too expensive for small businesses to buy unless it came with a franchise to sell ice cream. So, he sold the unit at the manufactured cost to store operators and made his profits on the soft-serve product that was produced by the machines. By June 1953, Tastee-Freez had grown to a national chain with over 600 outlets. *Author Collection/Courtesy Tastee-Freez*

By the 1960s, the Steak n' Shake chain of drive-in restaurants was well known throughout the Midwest and even Florida. Gus Belt's original formula for serving up a complete meal of Steakburger and hand-dipped milkshake became a roadside standard. "It's a Meal" was the company's popular slogan. Because the customer could see the order being prepared, "In Sight It Must Be Right" joined it as car-dining catch-phrase. *Courtesy Steak n' Shake, Inc.*

oring techniques that his company artists used to retouch scenes in order to make them appear more appealing.

Equally colorful and just as collectible are imprinted matchbooks. Back during the heyday of the drive-ins, some of the major cigarette brands actually promoted the idea that smoking was good for you. Since nearly everyone smoked or aspired to the habit, matchbooks were a really big deal—especially the free ones. Drive-ins didn't miss a beat and picked up on the mandate for after-dinner light-ups. Little match packs splashed with stylized art of every type abounded, filling what-not jars across America until the present day when they have finally reappeared for sale in flea markets and garage sales.

Without a doubt, the demand for drive-in related collectibles and artifacts is growing year by year. Now, even the mainstream marketing gurus are picking up on the craze. Mail-order catalogs with nostalgic sobriquets like "Back to the Fifties" feature a plethora of drive-in type items, all clever reproductions of the genuine articles. Replica window serving trays, complete with a faux milk shake and cheeseburger (made from wax), are some of the most requested items that are sold in the line.

For those with a hankering to relive the drive-in delights of their youth, all of the props required to cruise back into the past are available for purchase, including Coca-Cola button signs, neon clocks, scale models of once famous drive-ins, chrome-plated jukebox radios, and more. In the hearts and minds of motorists and collectors alike, the American drive-in restaurant is back.

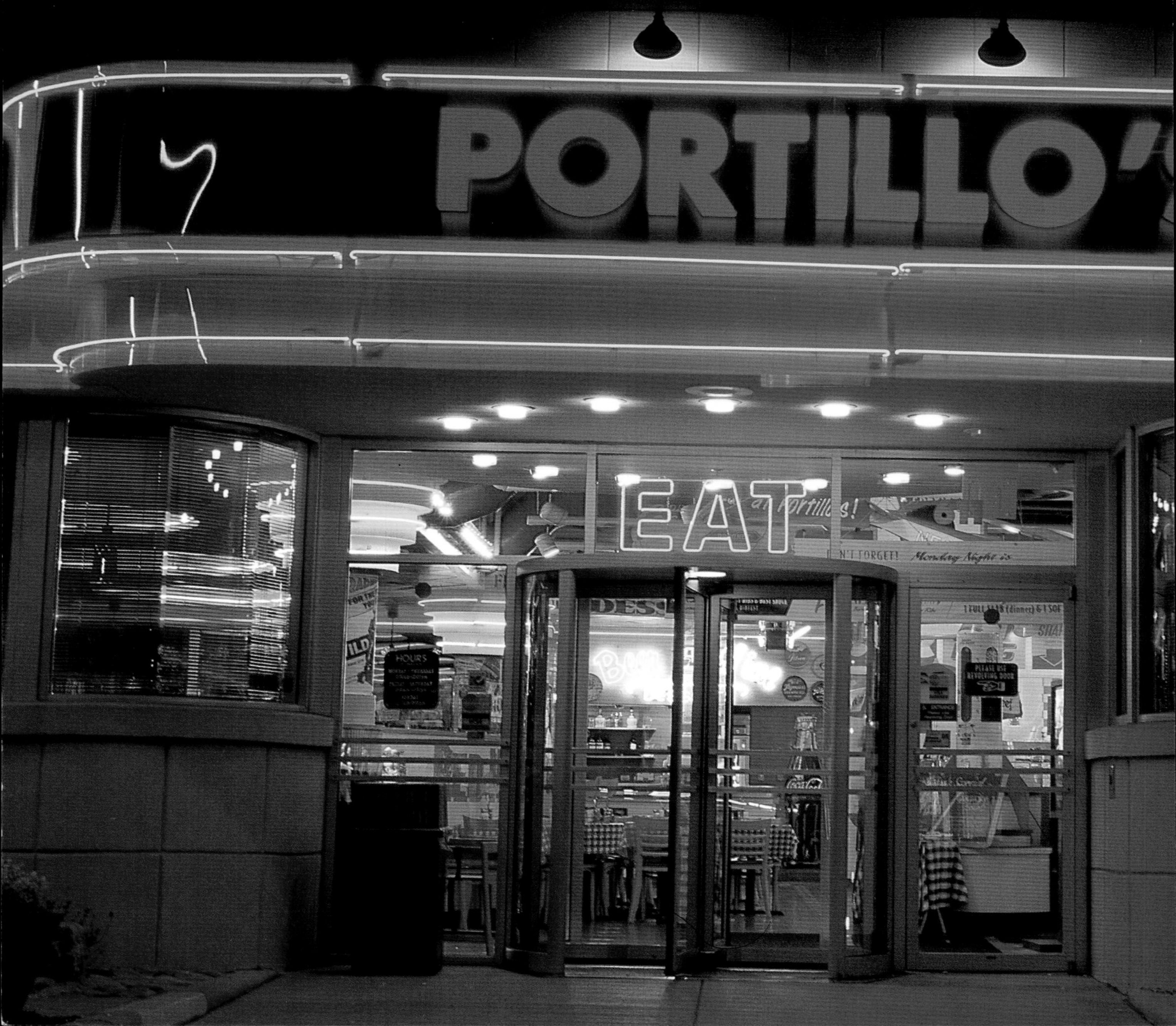
PORTILLO'S
EAT
HOURS
PLEASE USE REVOLVING DOOR

Chapter Eight

AMERICAN FAST FOOD FUTURES

When drive-in restaurant owner Troy Smith stopped for lunch at a Louisiana curb service spot back in 1954, he discovered that a clever cable talk-back system was being used to gather the customers' food orders. Energized with new enthusiasm, he returned to his hometown of Shawnee, Oklahoma, and immediately hired a local radio repairman to install one of the vacuum tube intercom systems at his own drive-in, a canopy-style eatery called the Top Hat.

Smith's radical order-taking device proved to be the first of its kind in the state. Inspired by the new rate at which orders could be taken and filled, Smith penned the catchy slogan "Service With the Speed of Sound" and made it the creed for his business. This was the beginning of the national Sonic drive-in chain and an entirely new direction

On Golf Road in Schaumburg, Illinois, Portillo's hot dogs employs every trick in the restaurant book to attract patrons beneath its colored bands of neon light. As a food that's easy to make and quite profitable, the hot dog and the many variations of it that may be made is a good choice for an operation that wishes to concentrate on making their specialty the best. *Howard Ande*

83

After Big Boy founder Robert Wian passed on, stewardship of the chunky mascot was assumed by a corporation. As hard as it was for loyal customers to believe, the bean-counters were contemplating his dismissal! After an unfavorable response from the public brought them to their senses, the ousting of the oversized mascot was put to a vote. Should the Big Boy stay or should he go? The answer came back a resounding yes: Americans liked the little butterball and wanted him to stay on as the company greeter and doorman. Nevertheless, some in the radical fringe weren't happy with the decision. The controversy came to a head in 1994 when bandits pilfered a 300-pound, 6-foot-high Big Boy from a Toledo, Ohio, restaurant. Showing little respect for the statue, they dismembered it with a hacksaw and dumped the pieces at Big Boy outlets in the surrounding area. Notes that were attached to the ragged fragments declared: "Big Boy is Dead." This Frisch's model on the old Lincoln Highway remains safe and sound, admired by happy diners. *Michael Karl Witzel*

for the future of curbside service that would one day be exemplified by a total lack of carhops.

With their practicality proven, the rush to install improved electronic ordering networks began in earnest. From 1951 until the end of that decade, a wide range of models were introduced. All the industry publications featured articles on the two-ways, introducing the modern restaurateurs to brands like Aut-O-Hop, Ordaphone, Fon-A-Chef, Serv-us-Fone, Teletray, Dine-A-Mike, TelAutograph, Dine-a-Com, Auto-Dine, and Electro-Hop. Each unit boasted features that included internally lit menus, record changers, dual speech amplifiers, and even the capability to play music. The future of fast food looked so bright that the drive-in owners were reaching for their shades.

While over the next 40 years the Sonic operation remained true to the carhop and speakerbox format, the majority of American fast food outlets chose to go in the direction of service minus the carhops. All too quickly, the energetic curb girl who had rocketed so quickly to icon status during the postwar era became almost extinct. Even the parking lanes where she once strutted her stuff were removed from the equation, replaced by a so-called "drive-thru" window where customers were required to pick up their order and exit to the street. A new-and-improved version of the remote-controlled ordering getup that was popularized by Troy Smith had usurped the responsibilities that were once appointed to a team of individuals.

Today, the modern-day menu board order taker has become the representative monolith for a rushed society that's in too much of a hurry. With a two-way intercom setup at its core and a chaotic patchwork quilt of backlit transparency display areas brimming with saturated, sumptuous-looking photographs of food and beverage choices, these automated service appliances have cleverly replaced all of the features that the drive-in restaurant-goer once held dear. The living, breathing carhop server and the one-on-one interaction that was enjoyed with her (or him) has been reduced to an impersonal two-way conversation via intercom.

The new ordering board has eliminated all need for the neat paper menus that could be held in the hand and studied at one's leisure. Presently, all possible combinations and variations of the basic hamburger and French fry pair-up are listed on the backlit order board, with the resulting conglomerate of information doing more to confuse the arrivee than to inform. Someone unfamiliar with a certain operation's particular food items would be hard-pressed to enjoy an unhurried selection. As cars honk their horns in the line behind, customers must scan and decipher all of the specially grouped meal-deals, combo platters, super-sized entrees, jumbo bags of grub, and other such price-conscious dinner promotions. The uncomplicated days of just pulling up and ordering a plain old cheeseburger are long gone.

At the same time, it's seldom that the consumer comes across one of these menu board gizmos that actually does what it is supposed to do. If the volume on the public address speaker (one that is usually mounted in close proximity to the eardrum and adjusted to a level that can compete with the noise level of a modern airport) isn't too

The Pig Stand sign seen at the San Antonio, Texas, restaurant has recently undergone a facelift. To compete with flashy eateries that line the roadway, neon tubes of searing purple were added to a color scheme that conjures up memories of the 1950s. Here the old sign is seen before the major facelift. *Michael Karl Witzel*

In Webb City, Missouri, the Bradbury Bishop Deli modeled its interior after the vintage soda fountains that were once common. Modern customers enter the nostalgic interior and are immediately transported to the days when milkshakes were made with ice cream and hamburgers were 100 percent beef. *Jim Ross*

Take a drive into any city today and the view is frighteningly similar: one after another, the very same fast food restaurants are lined up for the car consumer's convenience: Carl's, Wendy's, McDonalds, Burger King, Kentucky Fried Chicken, Whataburger, Taco Bell, Denny's, Spangles, Roy Rogers, Long John Silver's, and many more—they are all there on one strip of well-traveled road, signs shining brightly. *Michael Karl Witzel*

loud, it's way too low and filled with static—making it difficult to hear and understand the order taker cloistered inside the building. If the system happens to be operating at the correct volume level, then the worker inside is untrained in the subtle art of microphone communication.

It should come as no surprise that over the past 30 years, many a comedy sketch has been written and performed about this particular prob-

lem. It seems that no matter what amazing advancements in electronic intercommunication technology come down the pike, there's no relief in sight. During the 1970s the motoring fast food eaters of America nodded in affirmation and smiled quietly when the Jack-in-the-Box hamburger organization blew up their own speaker box clown in a series of hilarious television commercials. For a brief moment in the history of road-

Amid a panoply of competing restaurants and hamburger joints, it's difficult for even the most modern "theme restaurant" to withstand the pressure from the competition. Automotive-oriented restaurants like Hudson's appear—and then disappear—in the southern California region faster than you can dip two scoops and mix a double milkshake. *Kent Bash*

side food, one of the major players acknowledged the public's frustration and relieved some of the tension.

But a lot more has changed since the crazy days of the classic drive-in restaurant than just the intercom setup used to make faster orders. In the area of internal ordering protocols, the once whimsical banter that was exchanged between the order taker and the grill cook has been permanently eliminated. Where at one time a carhop might have called out loud to place her order for an "airdale" and a "stretch," special codes are now

Specialized hamburger outlets such as Rally's have given the big burger corporations a run for their money. Using nostalgic touches like glass brick, circular porthole windows, and neon lights, they appeal to the baby-boomer crowd that's interested in food quality and customer service. *Howard Ande*

being entered into a computer keyboard to signify the hot dog and Coke combination. At the twentieth-century hamburger bars of America, there's no room for folksy slang terms or friendly chatter.

Nowadays, the qualities of speed, efficiency, and customer turnover have been driven to the forefront. Instantaneously, the customer's selection is transmitted by bits and bytes along a maze of wiring to be displayed in contrasting type on a series of green cathode ray tubes that are mounted throughout the prep area. In this new Orwellian environment, duplicate television screens hang above the grill and checkout window, allowing all personnel to visually confirm at any

Detroit's Woodward Avenue is famous for the street machines that used to drive up the strip back in the 1950s and 1960s. Today, the essence of those times may be seen in restaurants along the entire length of this once hectic hot rod venue. The America Restaurant in Royal Oak recalls those drive-in cruising days with an unselfish use of classic materials. *Howard Ande*

given moment what "order is up" and where it is going. Despite these overly complex systems, human error is still a factor and mistakes in one's order are still made.

Still, it's not always the fault of these unsung fast food workers if an order is prepared incorrectly. The roadside playing field has changed dramatically since Richard and Maurice McDonald slapped

their first hamburger patty on the grill. These days, most of the roadside eateries are expanding their menus to gigantic proportions. Where at one time it was perfectly acceptable to sell only the standard food items, today's eateries are attempting to fulfill all of the consumers' desires. As a result, many unrelated entrees—including chef salads, pizza pies, dessert cakes, apple fritters, pita pocket

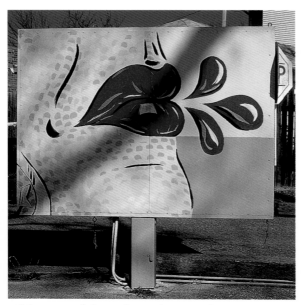

While most of the ordering boards of present-day fast food assume similar shapes, one may occasionally find a whimsical creation that's ready to take your food order. This big-lipped intercom board greets cars in Wichita, Kansas. *Michael Karl Witzel*

While deceptively simple on the exterior, the complexity of the sno-cone lies in its execution and creation. Yet, the elementary process of creating one is quite easy if you possess the knack for it. Dale's Sno-Cone stand in River Oaks, Texas, still follows the same classic method it has for decades. Owner Mamie Dale makes sure of it! Practicing her skillful methods every summer, Mamie established the criteria other stands are judged by and has proven that in the rare instance, the individually owned and operated stand can remain a viable force. *Michael Karl Witzel*

93

Checker's Drive-Ins are unique in today's fast food industry. Operations like this Chicago burger bar employ two separate drive-through lanes, making it possible to serve long lines of cars in times that the ordinary eatery can only imagine! What's next, three, four, five, or six lanes? *Howard Ande*

sandwiches, hoagies, chocolate chip cookies, ice cream cakes, sticky buns, egg rolls, and more—are being offered for sale to the modern consumer.

Instead of filling a niche market like the early pioneers of the fast food industry once did, roadside food chains are slowly reverting back to the drive-in days of the 1920s and 1930s when everything under the sun was served for general consumption. While on the surface this movement toward total inclusiveness might appear to be a boon for the traveling diner, the matter is up for debate. After all, certain entrees are specialties in themselves and aren't always conducive to the assembly-line processes that are employed by the roadside restaurant industry. Further expansion of the fast food menu board might water down the reputation attained from established food items.

However, there are a few select chain operations that are choosing to stick with the tried-and-true formula of selling just the staple items of favorite American road food. Rather than adding on exotic new meals, they concentrate on perfecting the mainline menu items and work to make them the best that they can be in terms of quality and taste. Efficient drive-in drive-through stands with names like Checkers, Burger Street, Rally's, and In–N–Out Burger are appearing on vacant parking lot spaces across the country, proving to the rest of the fast food industry that the basic hamburger and soft drink menu is more than enough to remain competitive on the roadside playing field.

INDEX